Newton's Laws: A Fairy Tale

Sarah Allen

Cover design by Vik Charlie.

Illustrated by Vladimir Djekic.

Except for: stars: © Ayeletkeshet | Dreamstime.com 69454131
and the chapter page tree image: Shutterstock

Book design by Sue Balcer.

Table of Contents

Newton's Laws: A Fairy Tale

Once upon a time, in a small village in the southwest corner of a tiny kingdom, there lived a farmer who had three sons. The eldest son was the strongest in the village. He regularly lifted whole oxen and set them back down again, just to show that he could. The middle son was the smartest in the village. He had fixed his father's irrigation system so that every plant got the optimal amount of water and, when he was only five, had redesigned the town's mill so that it ground twice as much wheat into flour every day.

The third and youngest son was called Kip and, as far as anyone could tell, he was not particularly special in any way. But he was happy. He loved his father and his brothers and the farm he grew up on, and he loved exploring. By the time he was twelve he knew every secret path and hideaway in the forest.

When it came time for the three sons to go out into the world, the father gave them each their inheritances. To the eldest he gave his farm, because he knew he would have the strength to work it. To the middle son, he gave his money, so that the boy could attend the university and become a wise scholar. And to Kip he gave all that he had left: an apple.

Kip, who was a small boy with mousy brown hair, looked at the apple. Then he looked at his eldest brother, who was already out plowing the fields. Then he looked at the middle brother, who was counting out gold coins and muttering calculations to himself.

Kip's father cleared his throat. "I know it's not much. It was all I had left. But I... I have a good feeling about this apple."

Kip would have liked to have been given the farm—everything from the way the trees moved in the wind to the secret glens of irises in the forest was familiar to him. But something in him yearned for the unknown. There might be more iris glens or snowy peaks or rushing rivers to discover. But if he had been given the farm, he might never have had the courage to leave it.

He would have liked to have been given the money, too. That way he could have bought a good travelling cloak and some food, and maybe a few nights at inns along the way. And... he couldn't really think of a consolation to that. Money would have been useful.

Kip didn't want his father to feel bad for only having an apple left to give him, so he smiled and hugged his father. "I have a good feeling about it, too," Kip said.

Kip packed his few possessions into a worn-out sack, waved goodbye to his father and brothers, and set off up the dusty road to seek his fortune.

The first night he slept in the woods on the side of the road. It was cold, and the ground was hard, but Kip lit a small fire and hummed a song to himself. He was hungry, but he didn't eat the apple. He was going to discover something great; he was sure of it.

The second night it rained, and Kip was unable to light a fire. He wrapped himself in his cloak and shivered, while the dark trees dripped water down on him.

The next day he had nothing at all to eat. He knew the apple would taste sweet and juicy, and he could have eaten every bit of it, even the core, but something told him to save it. The weather was cold enough that it wouldn't spoil quickly, and things would get better soon. He knew they would.

The next night it rained again and midway through the night the temperature dropped, and the water froze, and the rain changed to snow. Afraid he would freeze to death, Kip continued walking, barely able to keep to the road in the icy, moonless dark. He didn't even think about eating the apple. All he could think about was staying warm.

The next morning, when the sun came up and light filtered down in golden sunbeams, making the snow-covered branches sparkle, Kip came to a small village. He found a bakery and offered to sweep the floor in exchange for bread. Seeing how tired and cold Kip was, the baker agreed.

Kip swept and dusted, polished the windows until they were so clear they were almost invisible, cleaned the soot out of the chimney, and split wood. The baker was so impressed with his hard work that he gave Kip two whole loaves of bread and some cheese. Grateful, and warm now from the work, Kip went outside to wander the town and eat his food. He narrowly avoided being run over by a man galloping past on a black horse. Kip caught a glimpse of fine oiled leather and shining brass buckles as he stumbled back. When he had caught himself, he turned to watch the man ride away and realized it was a royal messenger.

A group of townspeople were clustered around something in the central square, talking excitedly.

"What is it?" Kip asked, approaching the group. A man turned and answered him gruffly.

"The king's dead." He touched his fist to his forehead and bowed his head in reverence. Kip did the same.

"How old is the new king?" a young boy asked.

"Didn't have any children," a woman answered, shaking her head.

"Who's going to be king, then?" Kip asked, feeling stupid. He took a bite of his bread.

The first man shrugged. "Doesn't say. The royal proclamation says that the council will decide."

Maybe it was because he'd just eaten the most delicious food he'd ever had, after being the closest to starving he'd ever been, but Kip felt a rush of excitement. It felt like a sign. Of what, he didn't know, but he knew what he needed to do next.

"Which road goes to the capital?" he asked the crowd, his mouth still full.

The man looked at him, and then chuckled. "This one here thinks he's going to be king."

"No, I don't," Kip mumbled, but the huddle of people laughed at him.

"You'd best be on your way, your majesty," the woman said, giving a mock bow. She pointed to the scroll tacked on the post. "It says they'll make their choice in nine days. The capital's at least a week's walk from here."

"Yes, hurry!" another man said, sniggering. "You've not much time, Your Highness!"

Blushing, Kip turned and walked away. The last bite of bread was heavy in his mouth, and he suddenly found he couldn't swallow.

"You're going the wrong way!" one of them called out.

Kip choked the last bite of bread down and kept walking. He listened for sounds of them following him, but their laughter faded away. When he could no longer hear them, he doubled back, turning down a side street and skirting the main square, finding his way to the other end of the village, where a sign pointed the way to the capital.

Kip walked alone for two more days, following the road as it wound into the mountains. He ate the last of his food on the second day. Again, he considered eating the apple—what good was it going to do him anyway? But something kept him back.

That afternoon another black horseman galloped past, a leather satchel at his side. More announcements from the capital, Kip thought, excitement swelling in his chest. He wondered what they were.

The next day Kip continued to walk, although now he was beginning to feel slightly faint from hunger. Which was why he thought he might be hallucinating when, late in the afternoon, he came across an old man lying face up in the middle of the road.

He was tall and skinny, with bony, hairy ankles that protruded from his midnight blue robes. His long white beard had

blown back in the wind and covered his face. Several feet away, his pointed blue wizard's hat sat in a mud puddle.

"Are you all right, sir?" Kip asked.

The man grunted, sucked in some of his beard, and, coughing, spat it back out. His hands scrabbled at his face, brushing the white hairs away and smoothing them down his belly. His bright blue eyes snapped open, and he glared at Kip.

"Fine, lad. Perfectly fine." He closed his eyes again.

Kip considered this. He had been raised to respect his elders, and it seemed wrong to leave such an old man lying in the middle of the road. He went over, picked up the hat, and wiped as much mud as he could off it.

"You dropped this, sir," he said, holding the hat out to the man.

The man glared at his hat, then at Kip, but he took the hat and stuffed it back onto his head.

"Do you need anything?" Kip asked.

The man sighed. "Can't a person have a moment's peace? I was trying to work something out, and I almost had it, until you came along."

Kip thought that if one wanted a moment's peace, they would be more likely to get it not by lying down in the middle of the road. "Sorry to bother you," was all he said. He could have walked away at that point. But something told him not to go just quite yet.

"Are you hungry?" Kip asked.

"Starving, actually," the man said, perking up.

And Kip knew what he'd been saving the apple for. He pulled it excitedly from his bag, shined it on his shirt, and held it out over the man's face so he could see it clearly.

"Here, have my apple," Kip said.

The man shrieked in terror, causing Kip to drop the apple, which hit the man full in the face as he tried to leap out of the way.

"Aaargh!" the man yelled, holding his head and stumbling away.

Kip moved after the old man, trying to help. "I'm so sorry, are you all right?"

The man stopped groaning and, still rubbing his face, looked at Kip. "It's not your fault. Apples have it out for me."

"Oh, um..." Kip tried to think of something sympathetic to say. Maybe... Apples could be that way? He'd always had a hard time with cherries. He didn't think he could manage to say either of those with a straight face.

"As you can surely see, I am a wizard. I was cursed by my arch nemesis, Leibniz the Evil, two hundred years ago," the man explained. He eyed the apple where it lay on the ground a few feet away. Kip followed his gaze, and to his complete shock, saw the apple roll over once, then twice, inching its way toward the wizard. He took a step back.

"It's fine, don't worry. It's got nothing against you. Just put it away, will you?"

Kip wasn't sure he wanted the apple anymore.

"At least get it away from me," the wizard said.

Reluctantly, Kip approached the apple, reached down, and picked it up. He half expected it to shake, or shock him,

or burn like acid, but it did none of those things. It was still an ordinary, if very shiny and red, apple. Not at all comfortable with the idea, he put the apple back in his pack. Still, it hadn't given him any trouble so far on his journey, and it was still his inheritance.

"Good. There you go," the wizard said. "I don't suppose you've got anything else?"

"No," Kip said. "Sorry."

The man examined him. Then he stuck out a bony hand. "Wizard Newton, the Great."

Kip had heard of several wizards, but not this one. "Kip." He shook the wizard's hand.

"Just Kip? Not Kip the Wise or Kip the Slightly Misshapen?"

Kip wondered if he was just pulling random titles out of midair, or if he thought either of those might apply. He himself didn't think he was very wise, and he hoped he wasn't misshapen.

"Just Kip."

"Pleased to make your acquaintance, Just Kip." He removed his hat and swept into a low bow. He shivered and looked around.

"It appears to be getting dark. Would you like to join me for dinner this evening?"

Kip knew better than to refuse an invitation from a wizard. Maybe the wizard might be able to conjure food for them. "Yes, thank you, sir."

Wizard Newton rubbed his hands together. "Wonderful. You make us a fire, and I will find us something to eat."

Kip found a dry place under some thick pines and gathered some twigs and pine needles. Then he cleared a space, although there was little danger of starting a forest fire in this weather, and piled up some small dry twigs and fir cones. They were drier than most of the wood, but they were still damp, and he struggled for several minutes to get a good blaze going. He had just begun adding larger twigs to the fire when Newton returned.

"I think I found a squirrel's cubby hole," Newton said, dropping a handful of moldy nuts next to the fire. "Roast those up for us, will you?"

Kip picked through the pile. A few dirty roots, which might have been turnips, some wilted leaves, and a handful of nuts. Kip's hunger and his politeness warred with his fear of food poisoning, but it was two against one, and he brushed the roots off, tied them to a stick, and attempted to roast them.

"Can't you call animals to us, or make plants flower out of season, or anything?" Kip asked. He was sure he'd heard of wizards doing that.

"Silly magic tricks. I'm not that kind of wizard."

"What kind of wizard are you?"

"Careful not to burn it, lad," Newton said, gesturing to the root, which was starting to smoke. Kip adjusted the root over the fire. "So, what are you doing out here on the road?"

Kip explained about his brothers and his father, and about seeing the sign in the village.

"Ah, so you're on your way to the Capital! Excellent, lad. I'm headed there, too. I was a great friend of the king, many years ago. It's a terrible thing, his passing. And terrible that he

hasn't left an heir. Those advisors of his are ridiculous. They'll tear up the kingdom with this fool plan of theirs."

"What plan?"

"Oh, you haven't heard?" Kip shook his head. "Well, they've decided how they'll pick the next king."

Kip perked up at this.

"Don't get too excited, lad. They said they thought about it and the thing a king needs to do is make laws. Laws that people will follow. So, they said that whoever wants to be king must make three laws and force the whole kingdom, everyone in it, to follow those three laws. Whoever is successful at the end of the week must present their three laws in the royal throne room for judgement."

Kip's enthusiasm diminished. How was anyone supposed to be able to do that? They'd need an army.

"Exactly," Newton said, causing Kip to jump. Had the wizard read his mind? "The worst part is, there is one person powerful enough to actually do that. Leibniz the Evil, my arch nemesis. He's a very powerful wizard, and completely unethical. If he takes over, he'll enslave everyone."

"But if he's the only one capable of succeeding at the challenge, how are we going to stop him?" Kip asked without thinking.

Newton looked at him. Then he patted him on the shoulder. "You're a good lad."

"Are you powerful enough to complete the challenge?"

Newton was examining the root. "What? No, I'm not that kind of wizard."

Kip wanted to ask, but he didn't think he'd get an answer.

Newton looked up and tapped the side of his nose. "But I've got an idea. That's why I was headed to the capital."

"Oh, what?"

"I'll just think of three things everybody already does and make those my laws."

Kip could see the logic of that, but he wondered if the advisors would allow it. It seemed rather against the spirit of the challenge. Also, what was something everyone already did? Breathing? He realized he was holding his breath. That wouldn't work. Unless they made a law about how often you had to breathe, or you would be put to death. But there were probably people who could hold their breath for a really long time. Also, maybe not everything needed to breathe.

"Yes, interesting thought, isn't it? I was just thinking things out when you came along, although I hadn't gotten very far. You seem like a likely lad. Will you help me? Assist me in this and I swear to you on my hat that I will make your fortune when I'm king."

Kip considered this, but only for a second. Something told him that this was exactly what he'd been waiting for. "Yes. I will help you."

"Wonderful!" The wizard reached out and clapped him on the back. Kip narrowly missed tumbling into the fire. "Let's eat!"

Newton's First Law

THE next morning, they continued down the road. The wizard was surprisingly quick, and Kip had to hurry to keep up. As they walked, the wizard kept up a constant stream of talk, rattling off idea after idea and shooting them back down again almost as quickly.

"Let's see . . . nothing can stay in the air . . . there's birds, though. But maybe nothing can stay in the air forever? Possibly true but difficult to prove, also Lieb'll just levitate something. That . . ." He glanced at Kip, cleared his throat, and went on. "Everything must rot. Nope, there's rocks."

"Rocks fall apart eventually," Kip volunteered.

"Not the same as rotting."

True. But there was something similar there. He thought about the baker's shop. "What about, like, everything gets more disorderly on its own?"

"Disorderly? What do you mean by that?"

"I mean, have you ever seen something clean itself?"

"Cats."

"But, would you say the cat is getting more orderly?"

"Hard to say, with cats."

"But, like, we're always having to clean, right? If something isn't cleaned, it gets dirtier, right? Always? Isn't that weird?"

"We're not looking for weird here, boy. We're looking for laws."

Kip sighed. He still thought it was a pretty good idea.

They continued walking, and he continued making suggestions, and the wizard continued shooting down both his own and Kip's ideas with equal rapidity.

Toward midmorning a group of soldiers came galloping up the road. Out in front was a tall woman with high leather boots and a red doublet. She swung out of the saddle as the horse pulled to a stop, whipping out a sword as she did so.

"Kneel in the name of Prince Eric!" she shouted.

Kip and the wizard stared at her.

"Kneel in the name of Prince Eric or I'll cut you down where you stand!" she cried.

Reluctantly, Kip and the wizard knelt.

"Good. Now, if anyone says, 'Prince Eric' in the future, you kneel."

"OK," Kip said, but he doubted he would. "Who is Prince Eric? I thought the king didn't have any children."

"Prince Eric was like a son to him. He was captain of the guards, and we guards think of him like a father. He's a good man, and the only choice for king."

"Why didn't the king appoint him his successor, then?" Kip asked.

"He died unexpectedly."

"Aren't you going to tell us to do something else?" Newton asked.

"What?"

"Well, it's three laws, not one. What are your other two?"

"Oh, no stealing. And no murdering."

Newton shrugged. "Those are pretty good."

"Prince Eric thinks so."

"Hard to get people to follow, though."

"Not for a true king."

Newton snorted, but Kip kind of admired the woman's devotion and Prince Eric's optimism.

"Now, let's practice," the woman said. "Get up."

She was still pointing the sword at them, so they got up.

"Prince Eric," she said, waving her sword in the air. Obediently, they got back down again.

"Very good." The woman smiled and hopped back on her horse. "Good day, then. Don't forget."

"Bye!" Kip called out, waving as they passed.

"Prince Eric, Prince Eric, Prince Eric," Newton muttered under his breath, stomping back down the road, and decidedly not kneeling.

The road wound higher into the mountains, cutting through dense forests and steep cliffs. They were walking along a narrow patch of road on the side of a fairly steep, rocky incline, and Newton was still muttering to himself, picking up rocks and tossing them, when he tripped and tumbled over the edge.

Kip lunged for him, trying to grab him before he went over, but he was too late. Newton bounced and slid, picking up speed as he went, until he ran headlong into a boulder, and crumpled to a heap on the ground.

Horrified, Kip leapt over the side, barely keeping his footing as he ran. He skidded to a halt and bent down over Newton, who was just starting to sit up, looking about himself dazedly.

"Oh my. Lucky that rock was here. Otherwise, I would have gone on forever." Suddenly, he leapt into the air. "That's it!"

"Er, are you OK?" Kip asked.

"What? No, I don't think so." Newton winced, but he began to climb the incline again, taking large strides with his bony legs. "But I've got it, lad!"

Kip stumbled after him. "Got what?"

"Don't you see? If I hadn't hit that rock I would have gone on forever!"

"Well, until you got to the bottom."

"Exactly!"

"What?"

"Everything must keep doing exactly what it's doing unless something forces it to change! I'll call those things 'forces.'" He grinned maniacally. They had reached the road again and stood, hands on hips, panting.

"But . . . but . . . that's just not true. I mean, nothing goes on forever."

"Oh, it would, if given the chance, my boy! Try me. Show me something that changes what it's doing, and I'll show you the thing forcing it to change. The force."

Kip picked up a rock and tossed it. It flew a few feet through the air, arcing up and then back down. When it hit the ground, it rolled a few inches and stopped.

"Easy!" Newton cried. He picked up the rock and rubbed it along the ground. "The ground is rough. The ground stops it!"

"So, you're saying that if the ground were smooth, it wouldn't have stopped?" Kip couldn't imagine this.

"Now you've got it!"

Kip didn't think he did.

"Look, let's find a rougher patch of ground. The rougher the patch, the sooner it'll stop." He tossed a rock onto some grass, and it stopped immediately.

A short way away, Newton found a frozen puddle. He tossed the rock onto that, and it skid several feet before stopping. "See? The smoother it is, the farther it goes. It's the roughness that's stopping it. If we had something perfectly smooth it wouldn't ever stop. Like when things are going through the air. They go a lot farther in the air than along the ground."

"That's true. But why do they always fall?"

Newton scratched his head. "Well, that's a tough one. Something must be pulling them back down. Otherwise, they wouldn't go down."

"But they're not attached to anything."

Newton waved his hands. "That doesn't matter."

"Doesn't it? If I want to move anything I have to touch it."

"Maybe it's something invisible. I don't know, lad, I was trying to work that one out earlier and gave up."

As they walked the rest of the afternoon, Kip looked around, trying to see if there was anything that didn't obey this law, but it was exactly as Newton had said. Birds turned by flapping their wings. Loose rocks tumbling down slopes stopped when they hit other rocks. And nothing that was stationary moved without something pushing or pulling on it first. It seemed like Newton might be right.

Newton's Second Law

THE next day they came to a narrow rope bridge. It was guarded by two men in Prince Eric's livery. One was very large, and the other was very short.

"Halt in the name of Prince Eric!" the large one said.

Not wanting any bother, Newton and Kip knelt.

"Have either of you murdered anyone or stolen anything today?" the small one asked.

"No, sir," Kip said, while Newton rolled his eyes and huffed.

"Good, you may be on your way, then."

Kip started for the bridge, but Newton wasn't moving. He was eyeing the two guards.

"You're not very evenly sized," he commented.

The large guard raised an eyebrow. "You can be on your way now, old man."

Newton ignored him. Instead, he addressed the small guard. "Everyone must attack you first, right?" He reached out and poked him. The guard narrowed his eyes.

"Newton, come on," Kip said.

"Just Kip, if you were going to attack one of these two guards, which one would you choose?"

"Neither of them," Kip said, alarmed. "Come on, let's go before they change their minds." He turned to the guards. "Sorry, sirs."

"Answer the question!" Newton roared.

Kip sighed. "The small one, I guess. But neither." He glanced at the guards, who were eyeing each other. "Definitely neither."

"And why is that?" Newton prodded.

"Because he's smaller."

"Exactly!" Newton smiled in a self-satisfied way, bowed deeply to the two confused guards, and swept past them onto the bridge.

"I have formulated my second law!" He waved his arms excitedly, and the bridge shook. Newton gave a little hop and chuckled as the bridge quaked. Kip closed his eyes and gripped the ropes on either side. "Don't you want to hear it?"

"Yes. But please keep going."

Newton continued making his way across. "Fine, fine. You're not as excited about this as you should be."

"Sorry."

There was a long pause. Finally, Kip said, "So, what's the law?"

"I'm glad you ask! The larger the object, the more force it takes to make it change what it's doing."

That made sense, but then he thought about the large bags of feathers he'd seen merchants carrying through his village. "What about big things that aren't very heavy, though?"

"Hmmm . . . Excellent point, lad, excellent point."

They'd reached the other side of the bridge, and Kip, feeling smart for having thought of something, was warming to the topic. "Isn't it really heavier things that are harder to move?"

"Yes, yes, you're right."

"Because, some things, even if they're the same size, don't weigh the same."

"Very true. It's not about size at all, it's about how much stuff is there. The more stuff, the harder it is to move. A bag of feathers is big, but it's not very compact. There's not as much stuff there, really."

"Why not just say 'heavier things' rather than 'more stuff'?"

"I don't know, that just feels wrong to me."

"But . . . aren't they the same? The more stuff there is, the heavier it is, right?"

"I can't think of a counter example, but it just feels like there might be a place where things don't weigh anything." He looked up into the sky and pondered for a few moments.

"What? But everything weighs something."

Newton shook himself. "Eh, you're right, I don't know what I was thinking."

"I think you're right, though." Kip said. "About the law, I mean. The more stuff in something, the more force it takes to make it change what it's doing."

Kip paused, then continued, "I don't like the word 'stuff', though." His middle brother had always told him it was a bad, non-descriptive word, and that it could always be replaced with a better one.

"Well, I don't want to call it weight."

"Fine. That guy was pretty massive. Why not call it mass?"

"Kip the Wordsmith!" Newton exclaimed. "Wonderful. Just wonderful. The more mass something has, the more force shall be required to move it. By law. By royal decree." He grinned.

Kip gave a half-skip and smiled.

That evening, as they sat by the fire that Kip had built and ate the random assortment of partially edible things Newton had gathered, Newton pulled out a large leather tome and began scribbling in it.

"What's that?" Kip asked.

"Oh, just something I'm working on," Newton muttered, sucking on his quill and squinting at the pages. It was so dark that Kip wondered how he could even see to write.

"Something magical?" He had yet to see Newton do any magic and was wondering what kind of wizard he was.

"Oh yes, very magical."

"Really? Can you tell me about it?" Kip leaned forward, forgetting the slimy mushroom he had been nibbling.

"Well, all right," Newton said, setting the book down and looking up. He got a misty look on his face. "I call it—" he held his hands up in the air and spread them wide in a grand gesture, like he was writing the word in the stars, "Calculus!"

"Oh." So, not a fire spell, or something that would conjure dragons. Or get them normal food.

Oblivious to Kip's disappointment, Newton held the book out to show him. Kip saw lines and lines of calculations in tiny, neat handwriting. He doubted even his middle brother could have made sense of it.

"What's it for?" he asked.

"Well," Newton said, rubbing his hands together. "Imagine you have a sphere, and the radius is changing. . ."

Kip tuned out after that. Newton spoke long into the night, his enthusiasm only growing as he described water flowing through culverts and collecting in ponds; men on horseback, one riding north at seven miles an hour and one riding east at twelve miles an hour; and farmers wanting to build pastures of maximum area with set amounts of fencing. It all sounded very complicated to Kip, but Newton was enraptured.

Newton's Third Law

THEY walked all the next day, talking about possible third laws, but nothing came to them. Several times they had to stop, kneel, and assure a representative of Prince Eric that they had neither stolen nor murdered recently. Neither said it, but the trial was only three days away and both were beginning to worry. Late in the afternoon, a strange pressure wave passed over them, and Newton went pale.

"What was that?" Kip asked, feeling faint. Before Newton could answer, another wave hit them, and this time, neither of them could move. For several seconds, Kip felt like he was held in an iron grip, unable to breathe. He was beginning to see stars when whatever it was released him and he collapsed, coughing, to the ground. Newton only stood, a grim expression on his face.

"Leibniz."

"What?" Kip looked around.

"Oh no, lad. He's not here. He's at the capital, I expect."

"But . . ."

"It appears he's working on a spell."

"Do you think everyone in the kingdom felt it?"

"Hard to say. We're close now."

"We'd better hurry, then!"

"No point in hurrying unless we figure out a third law."

Despite Newton's words, they both picked up the pace.

The next day they walked and walked, coming up with wilder and wilder ideas, but every idea had an exception. Nothing they could think of was true for everything, all the time. None of their ideas could be called a law.

Twice more during the afternoon they were stopped in their tracks, held captive by some invisible pressure. After each of these, they grew quieter, less certain of their plan.

The next afternoon, the day before the trial, they came around a bend in the road and there it was: the capital. It gleamed, a shining collection of towers and bright red roofs, with a white limestone castle rising in the middle.

Kip couldn't believe that something so beautiful had always been here, and he might never have seen it. His father had never seen it. Neither of his brothers had ever seen it. Only Kip.

Despite his hunger and his sore feet, Kip felt light and joyful just to be there. But they still needed a third law.

The road, which they'd had almost all to themselves for most of their journey, was now filled with traffic. Horses and

oxen pulled carts laden with crops towards the capital, and carts full of barrels and boxes and stacks of trade goods left it. Kip stared at everything. The clothes were strange and brightly colored, and the fabric was of such high quality. Even the farmers wore tunics that were nicer than his father's best wedding and funeral clothes.

Newton seemed oblivious to it all, striding along, looking down at the ground and mumbling to himself. But Kip felt his hope rising. Their laws were being followed everywhere, by everyone, and no one knew they were even doing it. Carts didn't start or stop without horses to pull them. Horses continued to pull on the carts to keep them moving because otherwise the rough ground would stop them.

The horse must pull exactly as much as the ground does, Kip thought. Also, that strange force that pulled everything down; it was working everywhere. Only, when things were on the ground, the ground held them up. Kip felt more and more confident. The only problem was that they needed a third law.

That night they slept in an inn, and Kip was ecstatic. He had real food, hot food, and an actual bed to sleep in. Both the food and the bed were richer than anything he'd ever had before. Newton was glum, though. They barely spoke through dinner. Kip tried to draw him out by asking about Calculus, but Newton only waved him off, muttering something into his stew.

The next morning, Kip awoke to see Newton sitting on the floor, staring up at the ceiling. There were dark circles under his eyes and a haunted expression on his face.

"It's no good, Kip. I haven't thought of a third law."

"Well, let's just use one of the ones we've already thought of. The first two are so good, maybe they won't pay as much attention to the third."

Newton put his head into his hands. "No. We know those don't work."

"Maybe they won't figure it out."

"Leibniz will. He'll be there. He'll point it out. I'm not going to go in with only two laws. It's a cheat, anyway. If I'm going to cheat, I should at least do it well."

The clock outside struck nine. The trial was at ten.

"Come on, we have to go," Kip said. "We'll think of something on the way."

"No, we won't," Newton said through his hands. "I can't take any more fruit attacks."

"What?"

"He'll just curse me again. Who knows what he'll do this time? I used to like fruit."

Kip jumped out of bed and grabbed Newton's arm. "Come on, we're not giving up now. Get up or I'm going without you. I'll take credit for all your work."

"Good, yes, you go without me," Newton said.

"Of course, I'm not going without you!" Kip said. "They're your ideas, you deserve to go."

"They're stupid ideas."

Exasperated, Kip dropped Newton's arm and looked around the room for inspiration. He saw his pack. He'd completely forgotten about the apple. He dug around until he found it, then pulled it out and held it up.

"If you don't come with me right now, I'm leaving this with you."

Newton looked up and flinched.

"In fact, I'm going to follow you around throwing apples at you for the rest of your life if you don't get up off the floor and come with me right now."

Newton glared at him, then at the apple, then he shuddered.

"Fine. Fine. Put it away."

"You'll come?"

"Fine. Yes." He looked down at his robes, which were mud splattered. "I can't go to the palace looking like this."

"You should have thought of that earlier," Kip said, stuffing the apple back into his pack. "Come on, get up. Let's go."

Very reluctantly, Newton got to his feet, sighed, looked down at his robes again, and then followed Kip out of the room.

Kip's mind raced as they made their way through the streets. He looked everywhere, at people pouring water out of buckets into culverts, at children carrying chickens and women pushing carts. But nothing occurred to him. He didn't give up hope, though. He might be the youngest son, and he wasn't as strong as his eldest brother or as smart as his middle brother, but he would think of something.

They reached the palace, where rows of trumpeters with golden trumpets lined the bridge across the moat. Newton's face was slightly grey now, but Kip smiled at everyone and took the wizard's hand as they walked into the grand entrance hall.

The walls and floors were marble, with gold filigree and glittering white statues everywhere. The hall was crowded with lords and ladies in silk, with gold and silver and jewels at their throats and around their wrists and embroidered into their gowns and tunics. In the middle was an open space where two men and a young girl stood.

The first was clearly Prince Eric. He wore black leather and carried a gleaming sword at his side. He had a noble expression and smiled at Kip and nodded politely to Newton as they came to join the group.

The second could only be the wizard Leibniz. He wore deep green robes embroidered with gold and encrusted with tiny, glittering rubies. He also nodded and smiled at Kip, but then ignored Newton, who ignored him back.

Next to Leibniz was a girl about Kip's age. She wore a simple green tunic, belted in leather, and her brown hair was tied back in a messy knot at the base of her neck. She grinned at him, and Kip blushed and looked away.

A hush came over the crowd as the clock struck ten, and a small gold door opened behind the empty throne. The five advisors filed out, each wearing a simple white robe and carrying a piece of parchment and quill.

They arrayed themselves in a line in front of the silent crowd. The middle one stepped forward, reading from his scroll.

"By royal decree, we will now hear and pass judgement on the laws made by the candidates for the crown. Prince Eric, you may begin."

At the sound of his name, everyone in the hall took a knee. Without thinking, Kip did, too. To his surprise, he saw Newton and the others kneeling, as well. Leibniz looked like he wanted

to laugh, and the girl at his side covered her mouth to suppress a giggle. Newton noticed what he was doing and, annoyed, tried to get back up again but slipped on the marble floor and fell over.

Prince Eric smiled, and several ladies in the room giggled. He turned to face the crowd.

"Thank you, ladies and gentlemen, for following the first of my three laws. It is the hardest to enforce, but a true king can do it. The second and third, of course, are that there will be no stealing and no murdering in thi—"

The girl darted forward, slipped a hand into Prince Eric's pocket, pulled out a coin and darted away, holding it up for all to see.

"Oops. Stealing." She grinned again.

"Give that back!" Prince Eric's face contorted, and he moved for her, his hand on his sword hilt. She laughed, leapt forward, and pushed him. He was so much larger, though, that her push barely impacted him, causing her to fly backwards instead, laughing and sliding away on the slick marble floor.

The idea hit Kip like a two-by-four to the back of the head. He knew what the third law should be.

Prince Eric was furious, but several noblemen had stepped forward to restrain him.

The advisor spoke again. "Thank you, Prince Eric. Unfortunately, it is clear you are unable to enforce your laws. You are disqualified."

Eric looked like he wanted to throw something at someone, but he settled for storming out of the hall. There was an uncomfortable silence before the advisor cleared his throat and continued. "Wizard Leibniz the Great, please give us your demonstration."

The great? But Kip thought Newton was Newton the Great.

Leibniz bowed. "If my assistant will return, please." The girl was playing with the coin, tossing it into the air and catching it. "Allora, please?" he said. She slipped the coin into her pocket and returned. They closed their eyes, each with their hands raised, and a crackling energy filled the hall. The pressure returned, stronger than it had been before, and for several seconds no one was able to move.

Afterwards, there was a lot of coughing and muttering and it took several minutes for the advisors to restore quiet.

"Most impressive, Wizard Leibniz. We shall check to determine whether the entire kingdom was affected, but so far it seems to be a success. And for your second law?"

Leibniz bowed. "Unfortunately, that was all I was able to accomplish. One law. Given more time I could likely make more, but, as you might guess, controlling an entire kingdom takes quite a lot of power."

They only had a single law? Kip saw Newton's head shoot up for the first time since they'd entered the hall. Hope gleamed in his eyes, and he started to look around frantically, clearly trying to think of something. Kip tugged on his robe to get his attention and when Newton looked at him, he mouthed "I've got it." Newton raised his eyebrows and looked at him in wonder.

"Wizard Newton the Lesser," the advisor said, and Newton cringed. "Do you have your laws?"

"I do," he said, straightening his back. He explained about the first law. "In summary," he said, "objects in motion will stay in motion, and objects at rest will stay at rest, unless acted upon by an unbalanced force." There was some confusion at this. No one had heard a law like this before, and no one could quite make out why one would want to have a law like that in the first place.

Everyone immediately set about trying to break this new law. Ladies kicked off their shoes and slid about the marble floors, men tossed their hats into the air, and the girl, Allora, ran around in the midst of the chaos pickpocketing people. The wizard Leibniz just smiled thinly. After a long while it was concluded that no one could break the law.

So, Newton explained about the second law. Various sized animals and boulders and bags of feathers were brought in to try to disprove it, but nowhere in the city could be found something that was more massive but easier to get moving or stop moving.

"Well, Wizard Newton the Lesser, you seem to have done well," the advisor said. "Although it does feel like you are cheating. What is the third law?"

Newton cleared his throat nervously and looked at Kip.

Kip felt himself blush as everyone in the hall turned to stare at him. He tried to think of how Newton would phrase it. "For every action, there is an equal and opposite reaction." There were a lot of confused looks, so he tried to put it into simpler terms. "You can't push on anything without it pushing back on you," Kip said. He accidentally caught Allora's eye, and she grinned at him. He blushed harder and looked away.

"If I push on a wall, I can feel it pushing back on me. If a horse pulls a cart, the cart pulls back on the horse."

"Wait, wait, wait," the advisor said. "If the cart pulls back on the horse, how does anything move?"

Kip thought for a moment. That was a good question. How did things move if every force had a balancing force? Then he had it. "They aren't pulling on the same thing," he said. "The horse pulls on the cart. So, it moves the cart. The cart, though, pulls on the horse. If both forces were pulling on the cart, then the cart wouldn't move, but one is on the cart and the other is on the horse. Forces that act on different objects don't balance each other out. Only forces on the same object balance each other out."

There was much arguing about this, but eventually most people agreed this was true, and the advisors were convinced. Newton looked at Kip and glowed with pride. Kip smiled, his heart pounding.

The advisors deliberated together for a few moments, then returned to address the crowd. "We are in something of a difficult position here," one said. "The task was to make three laws which everyone would have to follow. Leibniz has clearly succeeded in the spirit of the task. He has controlled the entire kingdom, which was the true nature of the task we set. However, he has only enforced a single law. Newton, on the other hand, has formed three laws, which, it is true, no one can break. However, we are of the opinion that these were somehow already laws before Newton informed us of them. So, in some ways, he has cheated. Discovered rather than made

laws. Still, is it not kingly to sense the true nature of the world? But, as king, one cannot just make laws about the way things already are."

The assembled nobles nodded wisely.

"And so, we have determined that the land must be ruled by a balance of the two. A king must know how to sense the truth of the world and work within that truth, but also a king must be able, in some instances, to impose his will upon the world. And so, we declare that Newton and Leibniz shall rule the kingdom jointly, combining their wisdom and strength."

Everyone in the hall smiled and nodded at this wise decision, except for Newton and Leibniz, who both looked like they'd been told they could eat nothing but moldy apples for the rest of their lives. Neither of them would look at the other.

Kip caught Allora's eye, and she nodded. She kicked Leibniz. Kip poked Newton.

"He's not evil at all, is he?" Kip whispered.

"Of course, he is. I told you about the apples."

"He doesn't seem that evil. How do you know him?"

"We were apprenticed together. He was always better than I was." Newton wouldn't look at him.

"But you've won, you've just bested him. Even with all his magic, you out-thought him."

Newton looked up. "I suppose I did, didn't I?"

"Yes, and look, you can be friends, can't you? You can forgive him?"

Leibniz was making his way over, looking mutinously at his apprentice, who continued kicking him every time he stopped.

"Newton," Liebniz said, in a forced voice, holding out his hand.

"Leibniz," Newton said coldly.

"Apologize," Allora hissed.

"I'm sorry," Liebniz said through gritted teeth. "About the apples."

There was silence.

Kip poked Newton. "Accept his apology."

"It's fine." Newton muttered this, but then he looked up and met Leibniz's eye. "Eh, it was pretty funny. Not a bad bit of magic."

Liebniz chuckled.

"Great!" Allora said, smiling at everyone. Kip smiled back.

Newton and Leibniz gingerly shook hands.

"Clever laws," Leibniz said finally.

"Not terrible magic," Newton said.

Leibniz smiled, and, reluctantly, Newton smiled, too.

They began to discuss each of their tricks more thoroughly, and Kip and Allora left them to it.

"He seems nice," Kip said, gesturing to Leibniz.

"Oh, he is. Just don't get him started on Calculus."

Kip stopped. "What?"

"It's this thing he's inventing." Allora picked at a thread on her sleeve.

Kip sincerely hoped that it wouldn't come up.

From then on, Leibniz and Newton ruled the kingdom together, assisted by their advisors, Kip and Allora. And, except for the Calculus thing, everyone lived happily ever after.

THE END

The Physics

(Some explanations of the concepts behind the story.)

NEWTON'S FIRST LAW

Newton's laws are widely known but also widely misunderstood. They seem obvious when you first hear them, but they actually contradict what most of us instinctively believe about the world.

Take the first law, sometimes stated like this: An object in motion will stay in motion, an object at rest will stay at rest, unless acted upon by an unbalanced force. Seems pretty basic, right?

Now ask yourself this question: When a ball is thrown, why does it come to a stop?

Most people instinctively say something like "because it runs out of force", which is wrong but fits with what we observe in the world. The truth, which Wizard Newton realized when he rolled down that hill, is the ball would have kept going forever except that gravity pulled it down, which meant it hit the ground. When it hit the ground, friction pulled it to a stop.

Newton's first law is often called The Law of Inertia. Inertia is the tendency of an object to keep doing what it's doing. Basically, something moving will keep moving unless some force stops it. The more mass something has, the more inertia it has.

We never see things move forever. This is because there's friction everywhere. If we grew up on space stations, we wouldn't have this misconception. We'd throw something, and it would sail through the air in a straight line at a constant speed until it hit something else. But here on earth, every single thing we throw or move or push or pull comes to a stop unless we keep pushing on it. Friction is so pervasive that we just don't notice it.

Let's take a moment to define some terms. There are a lot of words used in physics that are also everyday words, but they're sometimes used differently, or more precisely, in physics:

Force: A push or a pull. Some common ones include: Gravity, Friction, Tension, the Normal Force, Magnetic Forces, Electric Forces, Air Resistance, Applied Forces. I'll go into these more later, in the Problem Solving section.

Weight: Weight is a force. Weight is the pull of gravity on you. (So, weight is also often called the Gravitational Force or the Force Due to Gravity.) It is measured in 'Newtons' usually. Those are the standard units, but also if you say something weighs 12 pounds you are talking about weight. If you go to the moon, your weight will be 1/6 of what it is here on Earth. If you go into space, far enough away from any planet or moon or other massive body, you will be 'weightless,' meaning there is no force of gravity pulling on you.

Mass: How much stuff makes up an object. Not to be confused with weight. These are different, but very closely related, because the more mass something has, the heavier it will be (ie. the more weight it will have). Mass is measured in kilograms. In the everyday world, we often talk about kilograms as if they are a unit of weight, but that's actually not right. Kilograms are a unit of mass, a unit of how much stuff makes something up. Weight, on the other hand, is a measurement of how much the earth is pulling down on us. It's how heavy we are. If we go into outer space, we would be weightless, but we would still have mass. If we went to the moon, we would have less weight (because the moon pulls on us less,) but our mass would be the same. An object's mass is the same no matter where it is. To change the mass of an object, you must break a piece of it off or stick another piece on.

Velocity: How fast you're going. You can think of this as speed. It's basically speed, only it also includes direction. So, you can change your velocity by turning, even if you're still going the same speed.

Acceleration: How quickly your velocity is changing. Imagine you're driving at 30 mph. Then you push on the gas pedal and start speeding up. That feeling that you feel? That's acceleration. I saw a post on the internet yesterday that said, 'velocity is a funny thing'. It had a picture of the earth, showing how incredibly fast the earth rotates, and some smiling people having dinner, oblivious to the speed. Then it had a picture of a car driving at 60 mph and a smiling, totally calm woman driving.

Then it had a picture of a kid going down a slide. It was labelled as 3 mph or something and showed the kid's horrified face. Reading this, I was like 'noooooo'. Because the reason that kid is upset isn't that he's being a baby about such a slow velocity, and it's also not that 'velocity is weird'. It's that he's accelerating. We feel acceleration. We don't feel speed. Driving 30 mph feels the same as driving 60 mph which feels the same as being stationary. We don't feel anything until we try to speed up, slow down, or turn.

Try This: The Maze Phet Simulation

To help you understand the differences between position, velocity, and acceleration, try playing around with this Phet Simulation by the University of Colorado called 'The Maze'. (Also, a note about links: if you're reading this on a kindle without an internet browser, you can open this book instead in the kindle cloud reader, or on the kindle app on a smartphone, and then you'll be able to follow the link. Alternatively, you can just search for Phet Maze and it should come up.)

The goal here is to get the ball to the point labelled 'finish'. You can do this by controlling either the position, velocity, or acceleration. To start, make sure you've selected 'practice' in the green area and 'r' in the yellow area. Here, 'r' means "position". Then, click play. You move the ball by clicking and dragging on the blue dot above where it says "Position" in the lower right-hand corner.

Once you've scored a goal using position, keep it on 'practice' and try it with velocity ('v') instead. Then try it with acceleration, 'a'. Then bump it up to Level 1 and do the same thing.

YouTube Videos:

Veritasium makes awesome videos about physics. I can't recommend them highly enough. Some good ones to watch at this point are:

The Difference Between Mass and Weight
Egg Experiment to Demonstrate Inertia

NEWTON'S SECOND LAW

Newton's second law is often stated simply as an equation:

$$F = MA$$

In the story, this was the law Newton thought of when they came to the bridge with the two differently sized guards. (Also, just as a side note, this is completely my invention. This is not how these laws were derived. The history of these laws is complicated. The idea of the second law was already known in Newton's time, but he proved some really amazing things with this idea combined with Calculus and Kepler's work on orbits and gravitation. I won't go into it here, because the math gets complicated—outside of the scope of this book—but you can read more about it here: https://science.howstuffworks.com/innovation/scientific-experiments/newton-law-of-motion2.htm., or in the history section of the Wikipedia page.)

The idea here is that acceleration is caused by force. The harder we push on something, the more it will accelerate. Meaning, the harder we push on something, the more it will speed up, slow down, or turn. Also, the more massive something is, the more force it takes to speed that thing up, slow it down, or make it turn.

Try This: Forces in One Dimension Simulation

A fun simulation to play around with to get a feel for forces and how they cause acceleration is Phet's Forces in One Dimension. You click on the object and drag to create a force on the object (caused by a little stick man).

Try creating a force, then let go and see what the object does. Try turning friction off and on (in the sidebar on the right.) Try objects of different mass. Which of the objects requires the most force? How quickly do they stop once you stop pushing on them?

NEWTON'S THIRD LAW

Newton's Third Law is also one of those things that seems obvious but is actually more complicated than it seems. It is usually stated as "For every action there is an equal and opposite reaction." This feels so commonplace, it almost sounds like "haste makes waste" or "a job worth doing is worth doing well". Let's talk about what it really means.

In the story, Allora pushes on Prince Eric and goes flying backwards. Anyone who has gone ice skating and pushed off against a wall or sat in a wheeled chair and pushed back from a table has an intuitive feel for this.

The classic misunderstanding of this law is stated in the story. How does anything move? If every force has an equal and opposite reaction force, why does anything move at all? Why don't these forces just cancel each other out?

The answer is that these forces don't cancel each other out because they don't act on the same object. When two equal forces act on the same object from opposite directions, they do of course cancel out. If you push on a box from one side and your friend pushes on the same box from the other side, if you push with the same amount of strength (force), then the box won't move. Your efforts will cancel out. That's not what Newton's Third Law is talking about.

Newton's Third Law is for what are called "action-reaction pairs". Meaning, basically, the push-back from whatever action you're taking. So, if you push a kid on a swing, you can feel the kid pushing back on you. If you kick a soccer ball, you can feel the force of the impact on your foot. If you're sitting in a wheeled office chair and you throw a heavy computer, you'll roll backwards. The two forces, the action-reaction pair, are often called "third law pairs".

This always reminds me of the joke "Chuck Norris doesn't do push-ups, he does earth-downs". Technically, we all do "earth-downs" when we do push-ups, it's just that the earth is so much bigger it doesn't notice. But we both experience the same amount of force.

YouTube Videos

Three Incorrect Laws of Motion
Best Film on Newton's Third Law. Ever.

NEWTON AND THE APPLE

The classic story of Newton is that he discovered gravity when an apple fell on his head. I always thought this was just a made-up story, but apparently Newton did say that he started thinking about gravity when he watched an apple fall. You can read this New Scientist Article: https://www.newscientist.com/blogs/culturelab/2010/01/newtons-apple-the-real-story.html for the details." I felt like any story I wrote about Newton had to include an apple in some way, and I thought it would be funny if instead of one apple falling on him, all apples fell on him.

Also, the "Newton" in this story is not meant to be a portrayal of the actual Newton. The real Newton was a genius, one of the greatest thinkers and scientists to ever live and gave humanity an incredible amount of knowledge. I hope I'm not doing him a disservice by giving his name to a genius but comically flawed fairy-tale wizard character. Here's some interesting reading about what he may have actually been like as a person: https://www.quora.com/What-was-Isaac-Newton-like-as-a-person.

This concludes the conceptual side of things. I hope you've enjoyed it! Next, for people who are interested in the math or who are taking a physics class, I'm including a section on problem solving. It assumes a basic familiarity with algebra. Feel free to skip it if you're not interested. You can still go on and try the Conceptual Practice Problems at the end if you'd like. If you enjoyed this book, check out the rest of the series on Amazon!

PROBLEM SOLVING

This section is for people who want to get into the math of doing calculations with Newton's Laws. Feel free to skip it if your main focus is the conceptual understanding. Also, in this section I'll assume a basic familiarity with Algebra. If you want to brush up on your math skills first but you're not sure where to start, check out Khan Academy's Algebra 1 Course Challenge.

Problem solving with Newton's Laws usually involves determining which forces are acting on an object and then using this to calculate the acceleration of the object. Sometimes we work backwards from the acceleration and determine the forces on an object from this.

The main tool we use for thinking about the various forces on an object is called the Free Body Diagram.

In a Free Body Diagram, we start by drawing a square to represent an object. Then we draw a dot in the middle, like this:

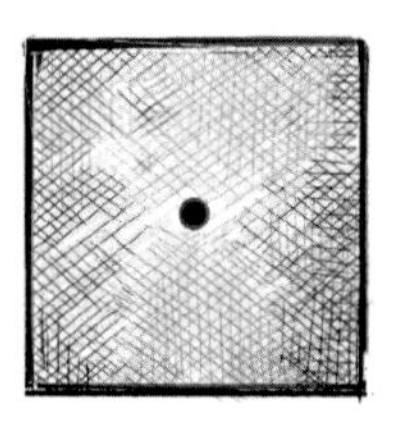

Then we draw all the forces acting on the object as arrows coming from the center of the object, like this:

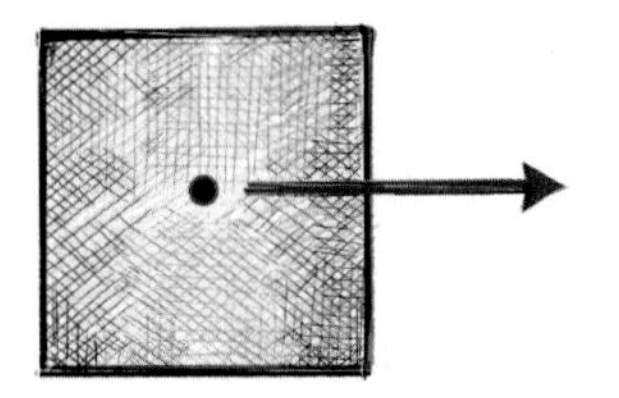

The length of the arrow represents the strength of the force. The longer the arrow, the stronger the force. It doesn't have to be super precise; there's no set length they have to be. It's more like, if you know one force is bigger than another force, you draw that arrow longer.

Now, let's go over some of the different types of forces that might be included in our diagram:

Gravity

The gravitational force is the weight of the object, or, how much the object is being pulled down by gravity. When we're on Earth we can use Newton's Second Law to calculate the force of gravity on an object. This is because gravity accelerates all objects equally.

"Acceleration" is basically how quickly an object's speed is changing. (Technically I shouldn't use the word speed, I should use the word velocity, but I'm not going to get into vectors here.) We measure speed in meters per second (m/s). Let's say I'm going 10 m/s but I'm speeding up. Two seconds later, I'm going 18 m/s. I sped up by 8 m/s, and I did that in 2 seconds. So, on average I got 4 m/s faster every second. That's acceleration.

We would call that an acceleration of 4 meters per second per second.

Note: the units for acceleration are usually written like this:

$$m/s^2$$

Going forward, though, I'm going to write it like this: m/s/s, which is read "meters per second per second." I'm choosing to write it this way for two reasons: 1. It's easier to type, and 2. I think it's clearer conceptually what m/s/s means. It's the change in the speed of an object. Meaning, it's how many meters per second (m/s) are added or subtracted every second.)

The force of gravity gets weaker as you get farther away from the Earth, or as you go higher, but around the surface of the Earth, the acceleration due to gravity is around 9.8 meters per second squared. If we want to know how much gravitational force this is, we can use Newton's Second Law:

$$F = MA$$

Here, m is the mass of the object and a is the acceleration of the object, which will be 9.8 m/s/s.

Force is measured in Newtons, which we abbreviate with a capital N. When we multiply kilograms by meters per second per second, we get Newtons.

If we had an object of mass 30 kg, we would find the force of gravity on it like this:

$$F_G = (30\ KG)*(9.8\ M/S/S)$$
$$F_G = 294\ N$$

Incidentally, we use the little subscript 'g' to indicate that it's the force of gravity on an object. Also, just so you know, 9.8 is often just written as 'g'. So, we often write the equation for the gravitational force as:

$$F_G = MG$$

This is because:

F = FORCE
G = GRAVITY
SO,
F_G = FORCE DUE TO GRAVITY

Here, m is the mass of the object in kilograms (not grams or any other unit. It needs to be in kilograms), and g is 9.8 m/s/s if we're on the surface of the Earth.

The Normal Force

The normal force is weirdly named. Don't think of it as "regular force". The word normal in math means perpendicular. The normal force is called that because it's a force that's always perpendicular to a surface. So, what is it? Well, it's sometimes called a "support force" because it's the force that holds things up, or supports them, when they're sitting on tables or floors. Basically, it's the force that surfaces exert on objects. For example, I have my cup of tea here. It's being pulled on by gravity as we speak, but luckily my desk is holding it up for me. That force that my desk is exerting on my cup is called the normal force.

Also, my chair is exerting a normal force on me, holding me up. My stack of Calvin and Hobbes books is exerting a normal force on my computer, holding it up. Usually, the normal force exactly equals the gravitational force, balancing it out.

Like this:

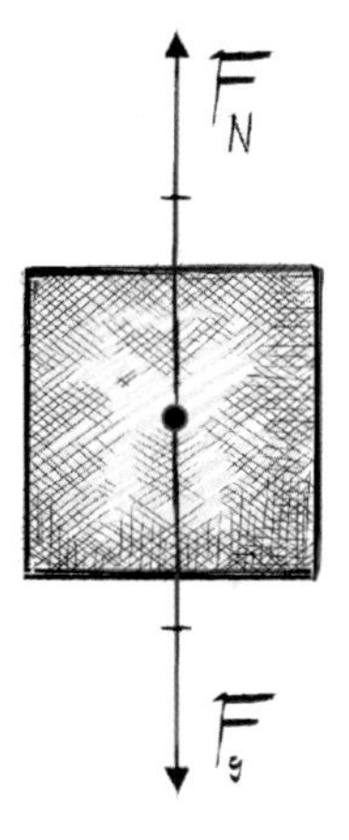

Here, we indicate that the two forces are the same strength in two ways. First, by drawing the arrows the same length, and second by drawing those little tick marks on the arrows. The tick marks indicate they are the same length. Notice also that we write the normal force as a capital F with a subscript N.

Friction

Friction is the force between two objects as they rub against one another. For example, when you push a box across a concrete floor, that's easier than pushing a box across a carpet because the concrete floor is smoother. The smoother the surface, the less friction. There are multiple types of friction, but I won't go into it here. For now, when I give you a practice problem, I'll just tell you how much friction force is being exerted. Friction is often indicated with a lowercase f. Sometimes a cursive f, like this:

Tension

Tension forces are the forces exerted by ropes and cables and things like that. When a rope pulls on an object, the rope gets stretched a little. The more stretched it is, the tauter it gets. This 'tightness' or stretch in the rope pulls on the object. It's like stretching a rubber band. The more you stretch a rubber band, the more it wants to pull back into itself.

We usually indicate tension forces like this:

F= force
T= tension

So,
F_T= tension force

Applied Forces

This is a broad category and includes things like people pushing on objects. These can be written a bunch of different ways, but here's how I like to write them:

Air Resistance

Air resistance is basically just the force of air molecules as they bounce off you. Imagine sticking your hand out the car window as it goes (hopefully someone else is driving). You can feel the force of air resistance on you. The faster you go, the stronger the air resistance is. I usually write this force like this:

Some people will write it as a frictional force. I don't like this approach, though, because I think it's misleading. Friction and air resistance work differently.

Electric and Magnetic Forces

These are super cool, but they're a huge topic and I'm not going to go into them much here. Electric forces are the forces that charges exert on each other, and magnetic forces are forces that moving charges exert on other moving charges. In general. It's complicated and totally fascinating.

Ok, now that we've got a set of forces to think about, let's draw some Free Body Diagrams. I'll give you some examples.

Free Body Diagram One

A book sitting on a table:

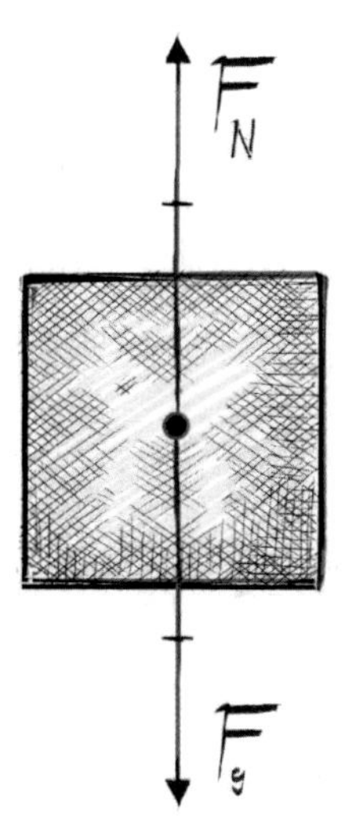

Notice that we've got the gravitational force because the book is presumably on Earth. Also, we've got the normal force because the table is holding it up. Now let's add some numbers to our diagram to make it quantitative. Let's say the mass of the book is 5 kg.

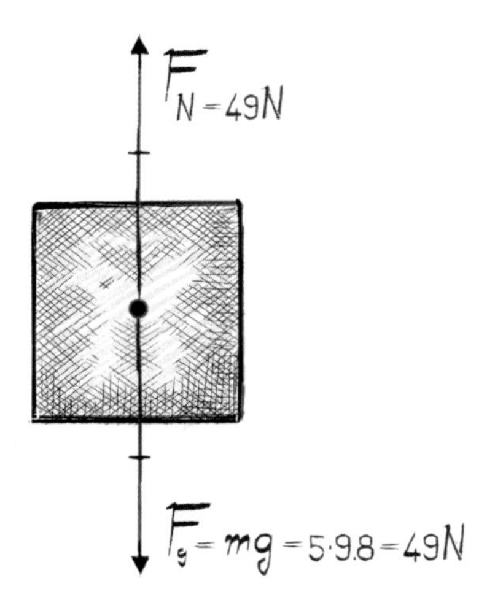

I calculated the gravitational force first, using the formula:

$$F_g = mg$$
$$= (5\ kg)(9.8\ m/s/s)$$
$$= 49\ N$$

and then I knew that the normal force was the same, so I just labelled it as 49 N, too.

Free Body Diagram Two

A book of 6 kg sliding over a frictionless surface at a constant speed of 4 m/s:

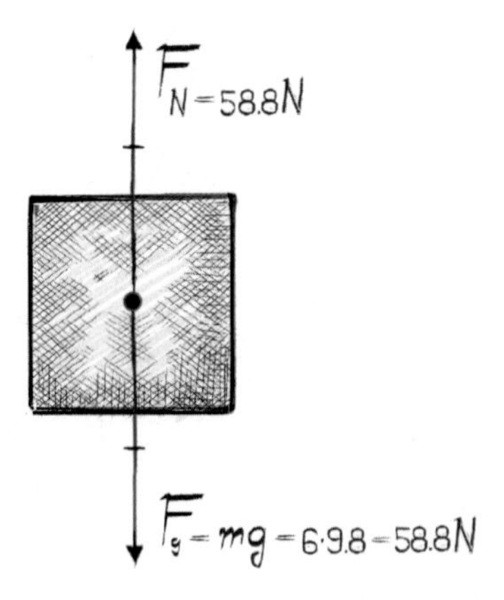

This is one that usually confuses people, because since the book is moving, they feel like there should be a force in the sideways direction. That's where Newton's First Law comes in! The book is going at a constant speed, so there doesn't need to be a force there. We only need forces to *change* motion.

Free Body Diagram Three

A box of 10 kg being pushed with a force of 50 N and experiencing a frictional force of 20 N:

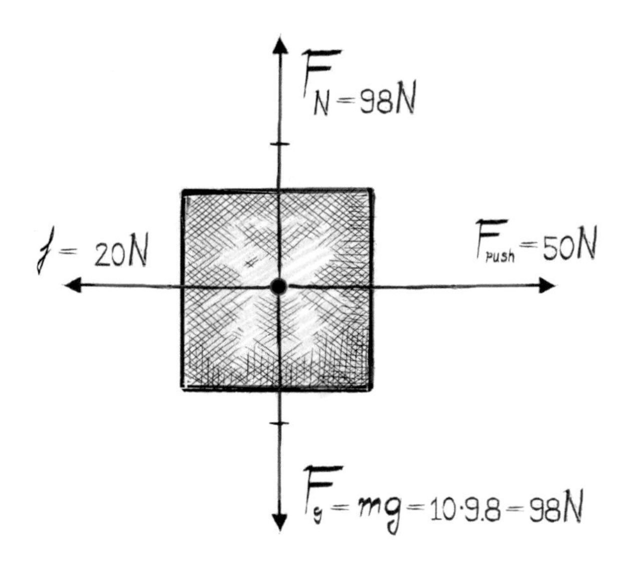

Again, I calculated the gravitational force first, then made the normal force equal to it, then drew on the applied force and the frictional force.

Free Body Diagram Four

A 0.2 kg baseball flying through the air:

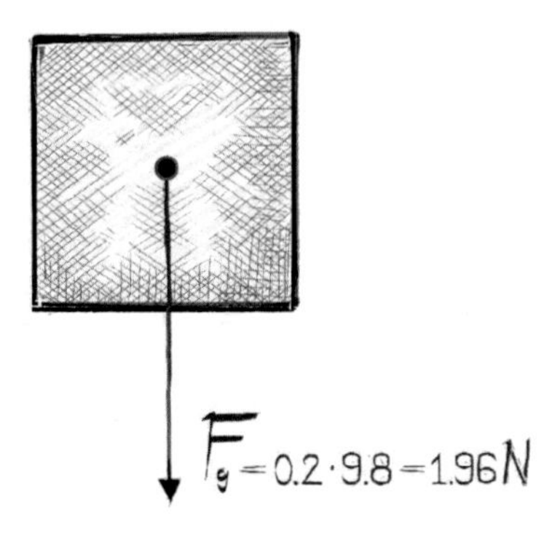

This one also trips people up. First off, in most cases, we ignore air resistance. This is for two reasons. The first reason is that it often doesn't exert very much force, so it wouldn't change our answer much, and the second reason is that it's really hard to calculate. It depends on a lot of tricky things about the shape of the object. So, since it doesn't have much of an effect anyway, we leave it out.

When something is flying, if we're ignoring air resistance, and if it doesn't have like wings or rockets or something making it go, then the only force acting on it is gravity. It doesn't matter whether it's going up or down or whatever. The Free Body Diagram always looks like this. That's the tricky thing to wrap your head around with Free Body Diagrams: they don't convey speed at all. They only show the forces on an object.

Ok, now let's talk about how to find the acceleration of an object.

In general, this is the process:

1. Draw the FBD with all the forces.
2. Find the net force.
3. Use F = ma to find the acceleration.

For example, imagine that you're pushing on a box to the right with 50 N, and your friend is pushing to the left with 40 N. The FBD looks like this:

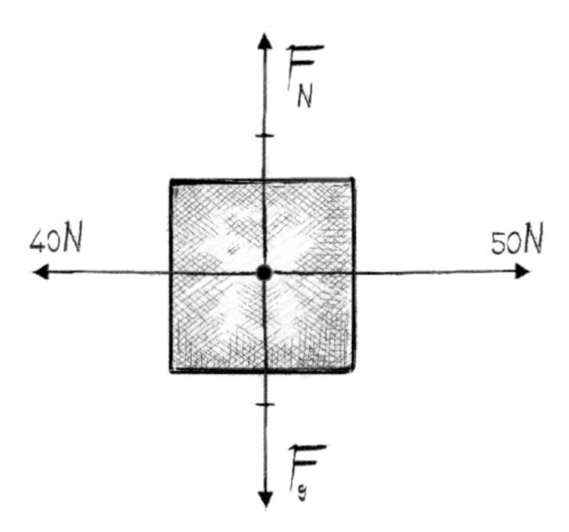

It makes sense that some of your force gets cancelled out, because you guys are working against one another. We call whatever is left over the 'net force'. In this case, the net force is 50 - 40 or 10 N.

We calculate the net force in the horizontal direction separately from the vertical direction. So, in this case, because the normal and the gravitational forces cancel out, the net force in the vertical direction is 0 N.

The net force is important because it's what causes the acceleration. (The force that gets cancelled out doesn't do anything to the box.) So, the net force is what we plug into our equation from Newton's Second Law. If the mass of the box above was 2 kg, we could find its acceleration like this:

$$F_{net} = ma$$
$$10 = 2 \cdot a$$
$$5\,m/s^2 = a$$

Now, let's move on to a tricky type of problem, the kind where you are given the acceleration of an object and want to find either its mass or one of the forces acting on it. In general, the process looks like this:

1. Draw a FBD, label all the forces you know.
2. Use F = ma to find the net force (assuming you have the mass)
3. Think about which forces should be bigger or smaller.
4. Either add or subtract to find the forces you don't know.

(This is a little vague, but I'll do an example next.)

Finding the Frictional Force

A 7 kg box is being pushed with a force of 50 N. It is accelerating at 4 m/s/s. What is the frictional force on the box?

First, I'll draw the FBD:

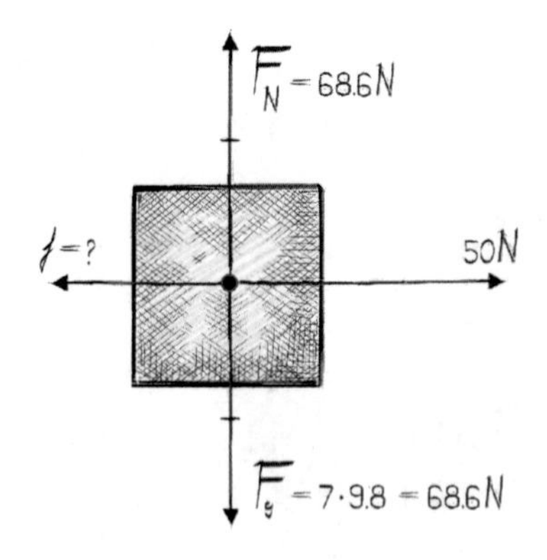

I knew that the frictional force would be smaller than the applied force, so I drew the arrow shorter.

Next, I'll use the acceleration and the mass to find the net force:

$$a = 4\,m/s^2$$

$$F_{net} = ma$$

$$= 7 \cdot 4$$

$$= 28N$$

Then I'll write an equation for the net force. The net force is the 50 N force minus the friction force, and I figured out above that the net force is 28 N.

$$50 - f = 28$$

$$+f \quad +f$$

$$50 = 28 + f$$

$$-28 \quad -28$$

$$\boxed{22N = f}$$

Ok, lastly, here's a classic problem called the elevator problem. Have you noticed that when you're riding in an elevator you feel lighter or heavier sometimes? Well, we call that the 'apparent weight'. Your apparent weight is just the normal force of the floor pushing up on you. Because the way you feel your weight is by how much the floor is pushing up on you. So, to find your apparent weight, you can just calculate the normal force, in the same way as we did above.

Finding Apparent Weight

You are riding upwards in an elevator. The elevator is speeding up with an acceleration of 3 m/s/s. You have a mass of 60 kg. What is your apparent weight?

Ok. So, same as before, start by finding the net force:

$$\begin{aligned} F_{net} &= ma \\ &= 60 \cdot 3 \\ &= 180N \end{aligned}$$

Next, draw your FBD:

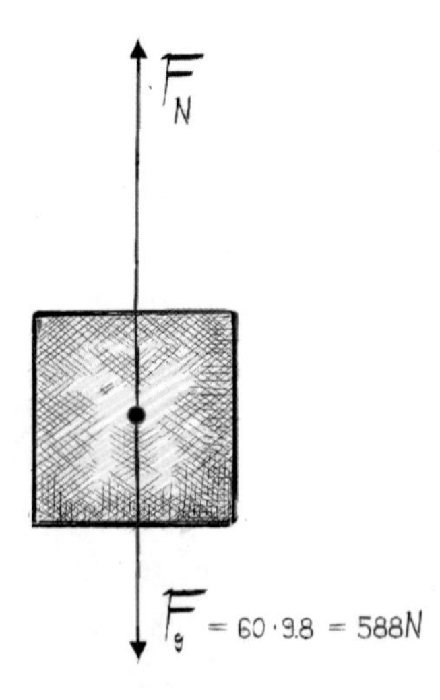

Here, I find it helpful to just imagine I'm in the elevator and ask myself whether I feel heavier or lighter. The elevator is going up and speeding up. I feel heavier. So, I make the normal force arrow longer. Also, I know that we're accelerating upwards, which means we need more force upwards to make the net force be up. So, now I can write this equation:

$$F_N - F_g = F_{net}$$

Now I'll just plug in numbers:

$$F_N - 588 = 180$$

And solve:

$$F_N = 768N$$

The normal force is 768 N, so my apparent weight is 768 N, too.

Thanks for Reading!

I hope you enjoyed this introduction to Newton's Laws! If you have any questions, you can email me at SarahAllenPhysics@gmail.com.

If you would like to hear about more cool experiments, YouTube videos, online simulations, and games, as well as get advance notice on new releases, freebies, and discounts on new books coming out, you can sign up for my newsletter on my website, www.MathwithSarah.com.

You can also find tons of free physics worksheets on my website.

The next sections are practice problems and answer keys. First are some conceptual questions, then there are some quantitative practice problems. Enjoy!

Practice

CONCEPTUAL QUESTIONS

(The answers are on page 77.)

1. A child and an adult are ice skating. They stand facing each other and push off from one another. Which one experiences more force, or do they experience the same force?
2. In the question above, which of them accelerates more?
3. If you kick a soccer ball with 10 N of force, how much force does it exert back on you?
4. Identify the third law pairs (the reaction forces) for each of these action forces:
   a. The gravitational force of the earth pulls a pinecone to the ground as it falls out of a tree.
   b. You push on a heavy trunk to slide it across the floor.
   c. You stand on a table. The table exerts a force on you to hold you up.
   d. You exert a force on a ball to push it into the air.
   e. You jump off a skateboard, pushing the skateboard back.
   f. A rocket shoots propellant out from its rockets.

g. You are sitting in a canoe and push an oar against the water.

5. An air hockey table uses air to hold the puck up off the surface of the table so that it doesn't experience much friction at all (almost none). If you had a giant air hockey table and you hit the puck to get it going, what would it do? Would it come to a stop?
6. In the previous question, if you hit the puck harder, what does that change about how the puck moves?
7. Here's a tricky one: Imagine you've got a tennis ball on a string, and you're swinging it around your head. What happens if the string breaks? How does the ball move? Draw a top-view sketch of what you think it will do.
8. Here's another tricky one: If you were in a windowless box in outer-space, and you couldn't feel any acceleration or hear any sounds, would you be able to tell if you were stationary (not moving) or moving at a constant speed?
9. Ok, I guess these are all just going to be tricky: If you were in the same windowless box, but you didn't know where you were, would you be able to tell if you were floating somewhere in outer space or falling to earth?
10. The Moon and the Earth are both pulling on each other with their gravity. How does the force of the Moon on the Earth compare to the force of the Earth on the Moon?

QUANTITATIVE PRACTICE PROBLEMS

(Be sure to give units for all your answers, and directions for all vectors (forces, velocities, and accelerations.) You can find the answers in the next section.

1. Draw qualitative (just the arrows, no numbers) Free Body Diagrams for each of the following objects:
    a. A cup sitting on a table.
    b. A hockey puck sliding along at a constant speed (assume the ice is totally smooth and frictionless).
    c. A box being pushed along at a constant speed.
    d. A rock flying through the air, moving upwards.
    e. A rock flying through the air, moving downwards.
    f. A rock flying straight up through the air, right at the highest point of its motion, hanging motionless for an instant before falling back to earth again.
    g. A box being pushed by a person so that it speeds up.
    h. A box moving to the right but slowing down because of friction.
    i. A box moving to the left but speeding up because someone is pushing on it.
2. Draw quantitative (including numbers) Free Body Diagrams for each of the following objects:
    a. A box of mass 5 kg is sitting motionless on a table.

b. A box of mass 12 kg is being pushed by a force of 30N to the right but is motionless because there is too much friction.

c. A box of mass 40 kg is being pushed to the right with a force of 100 N and is moving at a constant speed.

d. A rock of mass 3 kg is flying upwards in the air, after being thrown.

e. A rock of mass 6 kg is falling through the air after having fallen from a cliff.

f. Two friends are pushing on a box of mass 50 kg. One friend pushes to the left with a force of 100N, the other friend pushes to the right with a force of 80 N. Ignore friction.

g. Two friends are pushing on a box of mass 8 kg. One friend pushes to the right with a force of 60 N, the other friend also pushes to the right with a force of 10 N. Ignore friction.

h. A skydiver of mass 60 kg has deployed her parachute. The force of air resistance is 100 N.

i. A box (50 kg) is sliding to the right. The force of friction on the box is 50 N.

3. Use the free body diagrams you drew for question 2 to find the net force in each of these situations:
   a. A box of mass 5 kg is sitting motionless on a table.

b. A box of mass 12 kg is being pushed by a force of 30N to the right but is motionless because there is too much friction.

c. A box of mass 40 kg is being pushed to the right with a force of 100 N and is moving at a constant speed.

d. A rock of mass 3 kg is flying upwards in the air, after being thrown.

e. A rock of mass 6 kg is falling through the air after having fallen from a cliff.

f. Two friends are pushing on a box of mass 50 kg. One friend pushes to the left with a force of 100N, the other friend pushes to the right with a force of 80 N. Ignore friction.

g. Two friends are pushing on a box of mass 8 kg. One friend pushes to the right with a force of 60 N, the other friend also pushes to the right with a force of 10 N. Ignore friction.

h. A skydiver of mass 60 kg has deployed her parachute. The force of air resistance is 100 N.

i. A box (50 kg) is sliding to the right. The force of friction on the box is 50 N.

4. Use the net forces you found in question 3 to find the accelerations of these objects in each of these situations:
   a. A box of mass 5 kg is sitting motionless on a table.

b. A box of mass 12 kg is being pushed by a force of 30N to the right but is motionless because there is too much friction.

c. A box of mass 40 kg is being pushed to the right with a force of 100 N and is moving at a constant speed.

d. A rock of mass 3 kg is flying upwards in the air, after being thrown.

e. A rock of mass 6 kg is falling through the air after having fallen from a cliff.

f. Two friends are pushing on a box of mass 50 kg. One friend pushes to the left with a force of 100N, the other friend pushes to the right with a force of 80 N. Ignore friction.

g. Two friends are pushing on a box of mass 8 kg. One friend pushes to the right with a force of 60 N, the other friend also pushes to the right with a force of 10 N. Ignore friction.

h. A skydiver of mass 60 kg has deployed her parachute. The force of air resistance is 100 N.

i. A box (50 kg) is sliding to the right. The force of friction on the box is 50 N.

5. A child of mass 30 kg is sliding across a wood floor in his socks. He experiences a frictional force of 40 N. What is his acceleration?

6. A canoe of mass 200 kg is experiencing a friction force of 50 N from the water. What is its acceleration?

7. A rock of mass 40 kg is falling through the air. What is its acceleration? (Ignore air resistance).
8. A person of mass 70 kg is riding upwards in an elevator travelling at a constant speed. What is the normal force on them?
9. A person of mass 70 kg is riding upwards in an elevator, accelerating upwards at 1 m/s/s. What is their apparent weight?
10. A person of mass 70 kg is riding upwards in an elevator, accelerating downwards at 2 m/s/s. What is their apparent weight?
11. A person of mass 70 kg is riding downwards in an elevator. The elevator is slowing down at 5 m/s/s. What is their apparent weight?
12. A person of mass 70 kg is riding downwards in an elevator. The elevator is speeding up at 2 m/s/s. What is their apparent weight?
13. A person of mass 70 kg is riding upwards in an elevator, and the elevator is speeding up. Their apparent weight is 900 N. What is their acceleration?
14. A person of mass 70 kg is riding upwards in an elevator and the elevator is slowing down. Their apparent weight is 400 N. What is their acceleration?
15. A box of mass 30 kg is being pushed with a force of 500 N and is acceleration at 2 m/s/s. How much friction is acting on the box?

16. A sled of mass 50 kg is being pulled by a force of 60 N and is travelling at a constant speed. How much friction force is there on the sled?
17. A ball of mass 10 kg is flying through the air. What is its acceleration?
18. A plane of mass 3,000 kg is flying at a constant speed through the air. What is its acceleration?
19. What is the mass of a ball that is kicked with a force of 30 N and accelerates at 10 m/s/s?
20. What is the mass of a box that is pushed to the right with a force of 60 N and experiences a frictional force of 10 N and accelerates to the right at 3 m/s/s?

CONCEPTUAL QUESTIONS ANSWER KEY

1. They experience the same force (by Newton's Third Law). This is surprising because it feels intuitively like the adult should have been pushing harder. But imagine the adult pushing against a wall. However much the adult pushes on the wall, that is how much of a push they'll feel back from the wall. It's the same with the child. However much the adult pushes on the child, that's the amount of force they'll feel back. The child helps, too. It's like they create the force together. If the kid just let their arms go limp, it would be hard for the adult to push at all.
2. The child accelerates more because they are smaller (have less mass).
3. 10 N (by Newton's Third Law).
4. Answers:
   a. The pinecone pulls up on the earth as it falls.
   b. The trunk pushes back on you with the same force.
   c. You push down on the table.
   d. The ball exerts a force back on you as you throw it.
   e. The skateboard pushes you forwards with the same force.
   f. The propellant pushes the rocket forward.

   g. The water pushes forward on the oar (which pushes the boat forward, since you're holding the oar steady).

5. The puck travel in a straight line at a constant speed, until it hit a wall.
6. If you hit the puck harder, it just means it'll be going faster for that whole time.
7. The ball will move in a straight line, like this:

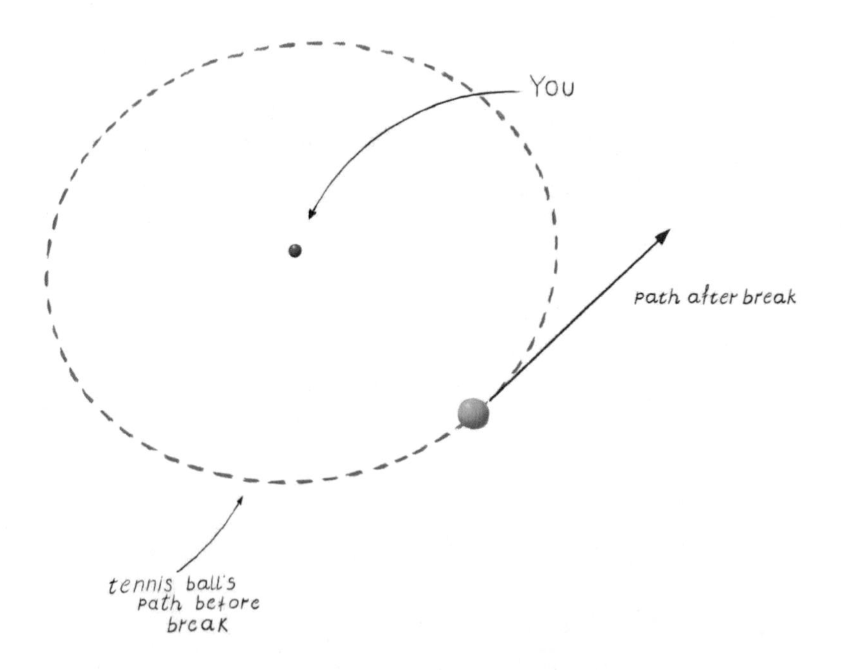

8. No, you wouldn't be able to tell. You might be stopped, or moving at 60 mph. It's a bit of a strange question, though, because if you're in space, what does speed even mean? What are we comparing our speed to? A passing

spaceship? The earth? The sun? All of those things are moving. So, it's hard to even say what our speed is, since we don't have some fixed reference point in the universe that we've all agreed is the "stationary" point.

9. No, again, you can't tell. This is why we can use those planes to simulate weightlessness. If you and the box are both falling to earth, then you're accelerating at the same rate. So, if you imagine being inside, you're still just floating inside that box. You can push off the sides and the top and float in the middle of it because it's accelerating at the same rate you are. So you feel stationary compared to it. To make it make more sense, have you seen videos of people skydiving together? Notice how they all are kind of floating next to each other? If they could only look at each other they might think they were stationary.
10. They're the same. (Newton's Third Law).

QUANTITATIVE PRACTICE PROBLEMS ANSWER KEY

Back to Practice Problems.

Answers:

1. Free Body Diagrams:

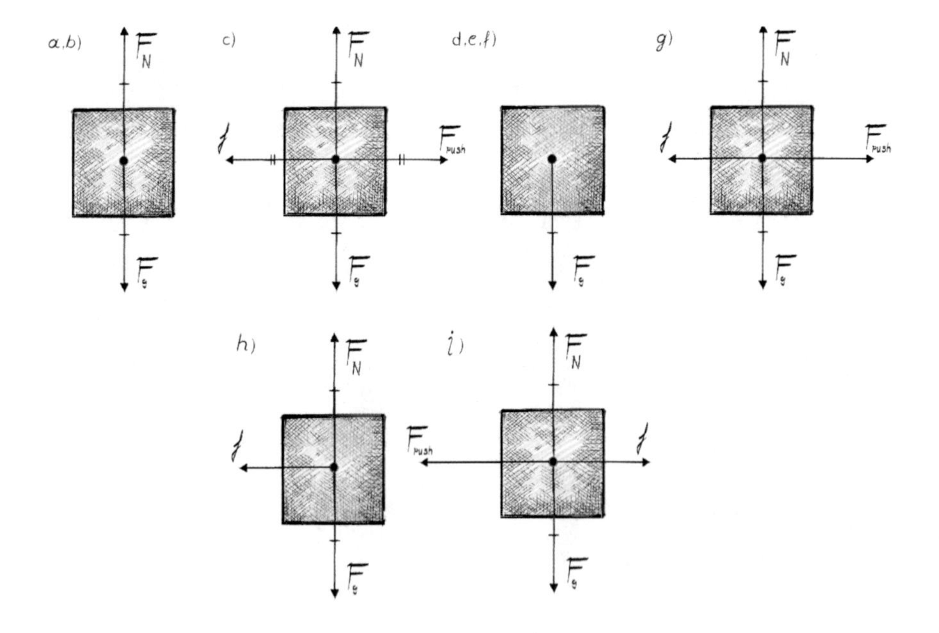

2. Quantitative free body diagrams:

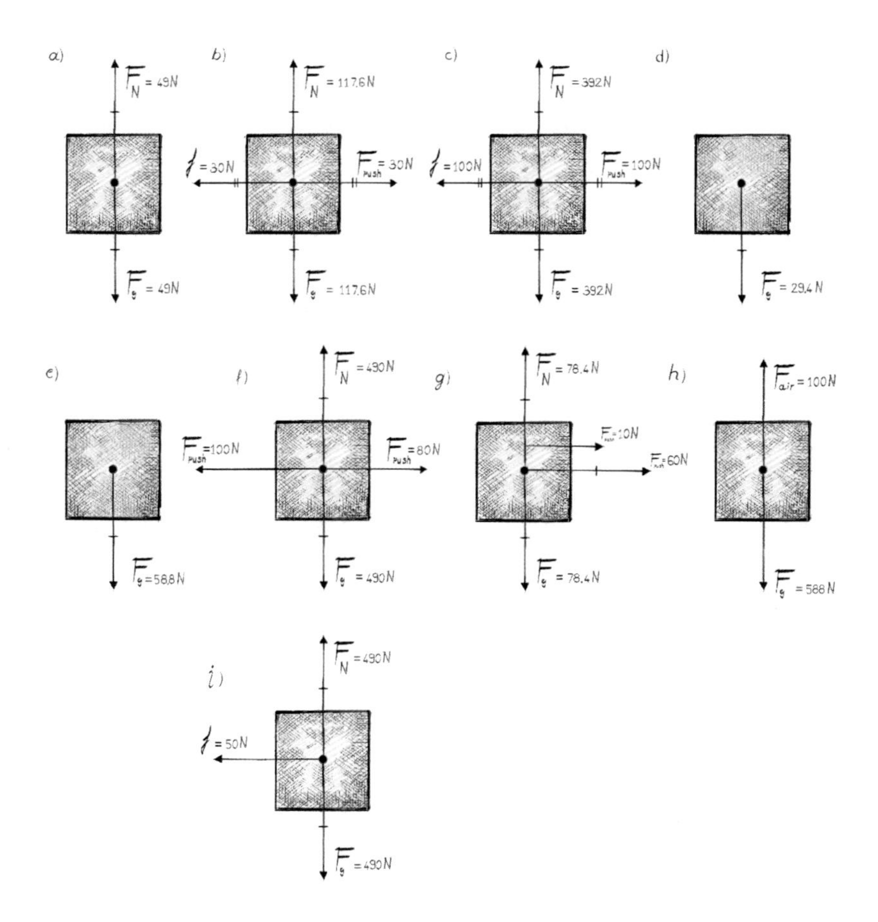

3. Net Force:
    a. 0 N
    b. 0 N
    c. 0 N

d. 29.4 N Down

e. 58.8 N Down

f. 20 N Left

g. 70 N Right

h. 488 N Down

i. 50 N Left

4. Acceleration:
   a. 0 m/s/s
   b. 0 m/s/s
   c. 0 m/s/s
   d. 9.8 m/s/s Down
   e. 9.8 m/s/s Down
   f. 0.4 m/s/s Left
   g. 8.75 m/s/s Right
   h. 1 m/s/s Left

5. -1.3 m/s/s (Instead of using the - sign to indicate direction, you could also say “opposite to the direction of travel of the boy.)
6. -0.25 m/s/s
7. 9.8 m/s/s Down
8. 686 N Up

9. 763 N Up
10. 546 N Up
11. 1036 N Up
12. 546 N Up
13. 3.06 m/s/s up
14. 4.09 m/s/s Down
15. -440 N (or 440 N opposite the direction of travel of the box.)
16. -60 N
17. -9.8 m/s/s
18. 0 m/s/s
19. 3 kg
20. 17 kg (rounded from 16.777...)

Made in United States
Orlando, FL
04 November 2024

53435426R00051